DISCARD

UNDERSTANDING EARTH'S SYSTEMS

Earth's Geosphere

Jenna Tolli

PowerKiDS press
New York

Published in 2019 by The Rosen Publishing Group, Inc.
29 East 21st Street, New York, NY 10010

Copyright © 2019 by The Rosen Publishing Group, Inc.

All rights reserved. No part of this book may be reproduced in any form without permission in writing from the publisher, except by a reviewer.

First Edition

Editor: Elizabeth Krajnik
Book Design: Rachel Rising

Photo Credits: Cover, p. 1 © iStockphoto.com/Vershinin-M; pp. 3, 4, 6, 8, 10, 12, 14, 16, 18, 20, 22, 24, 26, 28, 30, 31, 32 (background) maximult/Shutterstock.com; p. 4 natrot/Shutterstock.com; p. 5 Olga Danylenko/Shutterstock.com; p. 7 © iStockphoto.com/Yuri_Arcurs; pp. 7, 11, 13, 19, 29 Sakarin Sawasdinaka/Shutterstock.com; p. 9 Naeblys/Shutterstock.com; p. 11 Peter Hermes Furian/Shutterstock.com; p. 13 Fotos593/Shutterstock.com; p. 15 sirtravelalot/Shutterstock.com; p. 17 (rock cycle diagram) Spencer Sutton/Science Source/Getty Images; p. 17 (igneous) kavring/Shutterstock.com; p. 17 (metamorphic) www.sandatlas.org/Shutterstock.com; p. 17 (sedimentary) Fokin Oleg/Shutterstock.com; p. 19 I love photo/Shutterstock.com; p. 21 Putu Artana/Shutterstock.com; p. 23 Robert Cernohlavek/Shutterstock.com; p. 24 Edwin Verin/Shutterstock.com; p. 25 photographyfirm/Shutterstock.com; p. 26 the808/Shutterstock.com; p. 27 omihay/Shutterstock.com; p. 28 Joe Belanger/Shutterstock.com; p. 29 Rudmer Zwerver/Shutterstock.com.

Library of Congress Cataloging-in-Publication Data

Names: Tolli, Jenna, author.
Title: Earth's geosphere / Jenna Tolli.
Description: New York : PowerKids Press, [2019] | Series: Understanding
    Earth's systems | Includes index.
Identifiers: LCCN 2018000067| ISBN 9781538329757 (library bound) | ISBN
    9781538329771 (pbk.) | ISBN 9781538329788 (6 pack)
Subjects: LCSH: Geodynamics-Juvenile literature. | Geology-Juvenile
    literature. | Earth (Planet)-Juvenile literature.
Classification: LCC QE501.25 .T65 2018 | DDC 551.1-dc23
LC record available at https://lccn.loc.gov/2018000067

Manufactured in the United States of America

CPSIA Compliance Information: Batch #CS18PK: For Further Information contact Rosen Publishing, New York, New York at 1-800-237-9932

Contents

WHAT IS THE GEOSPHERE? 4
THE SURFACE AND BEYOND! 6
THE HOTTEST PLACE IN EARTH 8
LITTLE SHIFTS AND BIG
    MOVEMENTS 10
WHAT WE SEE AND FEEL 12
THE ROCK CYCLE 14
STUDYING THE GEOSPHERE 18
THE GEOSPHERE'S RELATIVES . . . 20
HOW WE FIT IN 22
MATERIALS WE NEED 24
FUEL FROM THE EARTH 26
PROTECTING THE GEOSPHERE 28
THE GEOSPHERE AND THE FUTURE . 30
GLOSSARY 31
INDEX 32
WEBSITES 32

What Is the Geosphere?

Have you ever wondered what makes up the ground we walk on every day? It's the geosphere! The geosphere consists of all nonliving matter on Earth, including rocks, minerals, and soil. Earth's interior is also part of the geosphere. Even though we can only see what is on Earth's surface, there are layers under it that play an important role in Earth's many processes.

The geosphere is just one of Earth's systems. The atmosphere, the biosphere, and the hydrosphere are the other systems that make Earth one of a kind. These four systems work together and make it possible for us to live on Earth. It's important for us to study these systems so we can keep Earth a safe place to live.

The next time you're outside, take a look around. Almost everything surrounding you is part of one of Earth's systems.

SYSTEM CONNECTIONS

Geologists, or people who study Earth, estimate that our home planet and the other planets in the solar system are about 4.6 billion years old! We can learn about Earth's age by studying rocks and minerals.

The Surface and Beyond!

The crust is the solid outer layer of Earth. The mantle, outer core, and inner core are the layers below Earth's surface. Together, these layers make up the geosphere.

The crust includes the ground we walk on, mountains, and even the bottom of the ocean. Even though it has so many parts, the crust is actually the thinnest of Earth's layers. It's an average of about 19 miles (30 kilometers) thick. Many important processes take place in this layer.

The mantle is below the crust. It is about 1,800 miles (3,000 km) thick, making it the thickest layer of Earth's interior. The crust and the upper mantle make up the lithosphere. The asthenosphere, also a layer in the mantle, sits beneath the lithosphere. The lower layer of the mantle is closer to Earth's outer core and is hot and soft.

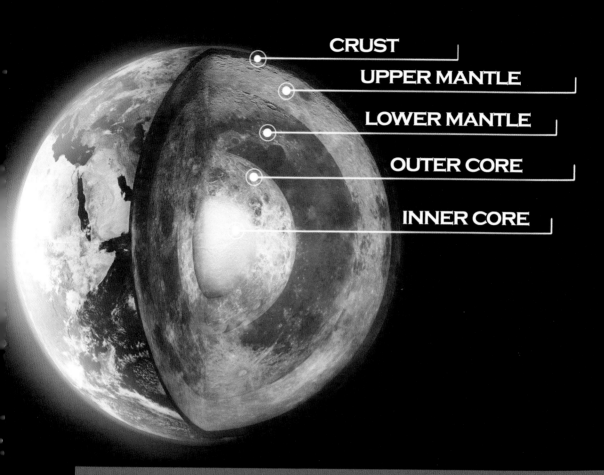

Even though we can only see Earth's crust, a lot is happening beneath our feet. Earth's layers have important roles in how our planet functions.

The Lithosphere

The crust and the uppermost part of the mantle are the solid, outermost parts of Earth. Together, these two parts are called the lithosphere and are about 60 miles (100 km) thick. The lithosphere is named after the Greek word *lithos*, which means "rocky." The processes that take place in the lithosphere, called plate tectonics, affect what we see and feel on Earth's surface.

The Hottest Place in Earth

Earth's outer core is below the crust and mantle. This layer is made mostly of iron, but it also contains some nickel and sulfur. It is also very hot—about 8,000 to 10,800°Fahrenheit (4,400 to 6,000°Celsius). You probably think of iron and nickel as solids. However, because the outer core is so hot, these **elements** are actually in their liquid state in the outer core. The outer core is about 1,400 miles (2,200 km) thick.

The inner core is the center of Earth, which is about 4,000 miles (6,400 km) beneath the surface of Earth. This is the hottest part of Earth's interior. This layer is made up of the same elements as the outer core. Even though the temperature of the inner core is high enough to melt the iron, great pressure prevents it from melting.

Scientists believe Earth's inner core is about 1,500 miles (2,400 km) across.

System Connections

The inner core is thought to be between 9,000 and 13,000°Fahrenheit (5,000 and 7,000°Celsius)!

Little Shifts and Big Movements

Earth's tectonic plates, which are part of the lithosphere, do not stay in one place. They move. The theory of plate tectonics explains how and why the plates move. It explains how mountains are made, why volcanoes erupt, and why earthquakes happen. It also explains how the surface of Earth has come to look the way it does today and what it used to look like in the past.

The solid plates of the lithosphere are like giant puzzle pieces. They all rest on top of the asthenosphere, which is partially molten. The solid plates can slide over the softer asthenosphere. The individual plates can diverge, or move away from each other. They can also converge, or move toward each other. Sometimes they slip past each other, side by side. Each year, tectonic plates move at a rate of one to two inches (2.5 to 5 cm).

We usually can't see fault lines, or the places where tectonic plates meet, but scientists can measure changes in the earth to find out where they are. One fault we can see parts of is the San Andreas Fault in California.

Tectonic Plates

The major, or largest, tectonic plates are named after Earth's continents, even though their boundaries sometimes stretch across other continents. The United States is part of North America, which is the third-largest continent on Earth. This means that we're on top of the North American tectonic plate, which is the second-largest tectonic plate on Earth.

What We See and Feel

When Earth's plates move, they can get stuck on rough patches of rock and lock together. Even though the edges are stuck, the plates keep slowly moving. Eventually, though, the edges get unstuck and this sudden movement is felt as an earthquake. Earthquakes can be large or small, and they can damage the communities nearby.

There are other effects from inside Earth that we can see and feel. Volcanoes are formed when tectonic plates either converge or diverge. Volcanoes can erupt when there is a crack in Earth's crust or in spots where the crust is weak. When magma, or really hot, molten rock, reaches Earth's surface after being released from Earth's mantle, it's called lava.

System Connections

The ancient city of Pompeii had a volcanic erruption almost 2,000 years ago. It destroyed the city. Today, we know how to better recognize and prepare for these events.

Volcanologists, or geologists who study the formation and behavior of volcanoes, keep track of warning signs and tell people if they need to get away from a volcano.

Measuring Earthquakes

Earthquakes are measured by their **magnitude**. We often hear news about earthquakes that damage buildings, roads, and even entire cities. Many more earthquakes actually occur that we don't feel, and only a small percentage of them cause damage. Scientists have estimated that there are 400,000 earthquakes per year that we can't even feel. These are typically called microearthquakes.

The Rock Cycle

Rocks are an important part of Earth because they make up part of the geosphere. Rocks contain different kinds of minerals. There are three types of rocks, and they form in different ways through a group of processes called the rock cycle. This makes the rocks look different from each other. It even changes what they're made of!

The three types of rock are igneous, sedimentary, and metamorphic. Each type of rock has specific characteristics, or qualities. We can see some of these qualities, but some we can't see. The rock cycle explains how these rocks form, break down, and change over time. The cycle includes weathering, erosion, deposition, compaction, melting, and cooling. The driving forces of the rock cycle are temperature, pressure, time, and environmental changes on Earth's surface.

SYSTEM CONNECTIONS

Plate tectonics are one reason why rocks change form as part of the rock cycle. Just as planet Earth is always moving, rocks are always moving too!

We can see different types of rocks everywhere. This is a picture of sedimentary rock in Arizona.

Igneous rocks are created when magma cools and solidifies. This type of rock forms at high temperatures in the asthenosphere. There, magma pushes toward the surface because it is less dense than the surrounding solid rocks. It may stay in that place or it may erupt through a volcano.

Sedimentary rocks are formed over time when water and wind break down rocks into much smaller pieces called sediment, which gets carried to rivers or oceans. Over many years, the sediment gets buried and makes up different layers, which form sedimentary rocks.

Metamorphic rocks are formed from existing rocks. The properties of rocks change when they're exposed to extreme pressure and heat. If the rock moves to the mantle where there is magma, it could get so hot that it melts and becomes igneous rock again.

> The rock cycle explains how rocks change over time. Any of the three rock types can turn into another rock type over many years.

System Connections

You might recognize granite from everyday household items such as countertops and tiles. The next time you recognize this rock, you can say it is an igneous rock!

The Rock Cycle

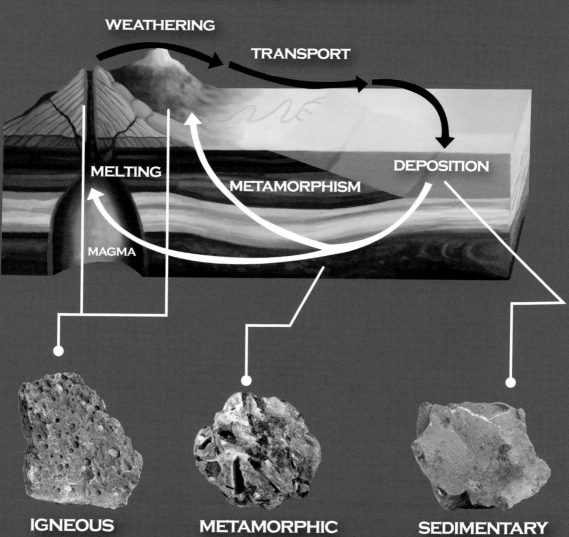

Studying the Geosphere

Since we can't actually see the inside of Earth, scientists need to use other methods to study how the geosphere works. Their findings help us understand how Earth has changed over time and how different processes in the geosphere can affect us.

When there is an earthquake, the ground beneath us vibrates. Scientists can measure these vibrations, which are called seismic waves, to give us clues about Earth's layers. By measuring the vibrations from different parts of Earth, we can see how the vibrations grow and change.

We also learn about Earth by studying fossils. We find fossils most often in sedimentary rock. They can tell us important information about ancient plants and animals. They show evidence of how Earth has changed over time.

The fossil of this prehistoric fish is found in sandstone, which is a sedimentary rock.

The Study of Earth

Geologists research materials such as rocks and minerals to learn more about Earth's interior and past life on Earth. Geologists also study processes like earthquakes and volcanoes to help us prepare for these events when they happen. When new buildings, bridges, or other structures are planned that might threaten Earth's geosphere, geologists help make sure it's safe for people to build there.

The Geosphere's Relatives

All of Earth's systems are related. The geosphere affects the other three systems, and the other systems affect the geosphere. For example, the geosphere is very important to the biosphere because all humans, plants, and animals live on the top layer of the geosphere—the crust. Without the geosphere, we wouldn't have a ground to walk on!

The hydrosphere interacts with the geosphere because most of Earth's water lies on or in the crust. Rain and other moving water, such as rivers and oceans, erode rocks. Wind, which is part of Earth's atmosphere, can also erode rocks. The atmosphere provides heat and energy that are needed for the different rock cycle processes to take place. When volcanoes erupt in the geosphere, ash and gases are released into the atmosphere.

Everything on Earth is connected. Earth's four systems work together all day, every day.

How We Fit In

We are always interacting with the geosphere, either by reacting to Earth's natural changes or by making our own changes to it when we need to.

There are some natural disasters, such as earthquakes and volcanic eruptions, that we can't prevent. Instead, we have to find ways to handle them to reduce harm to our communities. For example, cities that are in earthquake zones can construct buildings to resist the damage an earthquake can cause.

Sometimes we also need to make changes to the surface of the geosphere. When we need new bridges and tunnels, we might need to **excavate** rocks and soil to make enough room to build them. This can make the surface weaker. Engineers and geologists study Earth's surface before making changes and decide if the work is safe.

Sometimes we need to make new tunnels or bridges for cars or railroads. To build a new tunnel, we need to excavate existing rock, clay, or soil.

Materials We Need

Earth's geosphere provides us with many **natural resources**, such as oil, coal, sand, and rocks. We use oil and coal for energy. We use rocks for building. Soil is another important natural resource. We need soil to plant trees and food crops. The soil is also home to animals such as earthworms and insects. The top layer of soil helps plants grow because it has a lot of **nutrients**.

Minerals are another important resource from Earth that we use in our everyday lives. One mineral that is used a lot is quartz. Because it is a very hard and stable mineral, it's used to make things such as windowpanes, watches, and the eyeglass lenses. Another resource you may know is the element copper, which is used to make pennies and electrical wires.

STATUE OF LIBERTY

The top layer of soil has nutrients that plants need to grow. The ground becomes harder and less rich the deeper you dig.

System Connections

Did you know the Statue of Liberty is made of copper? Over the years, the copper has reacted to the oxygen in the atmosphere, a process known as oxidization. This has caused it to form the green coating that we see today.

Fuel from the Earth

Coal, crude oil, and natural gas are all fossil fuels that come from Earth. When the remains of plants and animals **decompose**, extreme heat and pressure turns them into what we call fossil fuels. This process takes millions of years.

We use fossil fuels for many important things. For example, coal is used in part of the process of creating electricity. Oil is **refined** to be used as fuel for cars, airplanes, and ships. Natural gas is used to heat our homes. Fossil fuels are still being formed today, but we use them up much faster than they're made. This is just another reason why we need to be mindful of how our actions today can affect Earth in the future.

Power plants burn fossil fuels to create energy. Coal-burning power plants like this one provide electricity for us, but they also cause pollution.

Protecting the Geosphere

We need to protect the geosphere in order to live on Earth, so we need to make sure we are able to sustain, or support, it over time.

As **technology** continues to advance, we may not need to use as many natural resources as we do today. For example, using more renewable energy sources such as solar energy or wind energy means we could reduce the amount of fossil fuels we use.

The geosphere is also where we have to dispose of waste. One thing we can do to help sustain the geosphere is to recycle and use **biodegradable** materials whenever we can. This will help decrease the amount of garbage added to **landfills**, which are part of the geosphere.

Solar panels collect energy from the sun and turn it into electricity, which we can use to light and heat our homes.

Using Earth's Energy

Geothermal energy is another type of renewable energy. It actually turns heat from inside Earth into energy that we can use. Unlike limited fossil fuels, heat from Earth is always being made, so we don't need to worry about it running out. Geothermal energy is also better for the environment than using fossil fuels.

The Geosphere and the Future

The geosphere is a very important part of Earth's systems. Each layer of Earth's interior has its own functions that help keep Earth cycling. The geosphere is where many continuous natural processes occur. These processes, such as the rock cycle, provide us with the resources we need to live on Earth.

There are some negative things that happen in the geosphere like earthquakes and volcanic eruptions as a result of these continuous processes. These events are out of our control. However, we have found ways to manage the damage they can create.

Some things are not out of our control. We should try our best to prevent damage to the geosphere's topmost layer by recycling, carefully considering new construction, and using renewable energy sources.

As we learn more about how Earth works, we will continue to find ways to protect the geosphere and the other systems of our planet.

Glossary

biodegradable: Capable of being slowly destroyed and broken down into very small parts by natural processes.

decompose: To break down into simpler parts or substances, especially by the action of living things such as bacteria and fungi.

element: Matter that's pure and has no other type of matter in it.

excavate: To uncover something by digging away and removing the ground that covers it.

landfill: A system of trash disposal in which waste is buried between layers of earth.

magnitude: A number that shows the power of an earthquake.

natural resource: Something, such as water, a mineral, or kind of animal, that is found in nature and is valuable to humans.

nutrient: Something taken in by a plant or animal that helps it grow and stay healthy.

refine: To remove unwanted substances from something and bring it to a pure state.

technology: A method that uses science to solve problems and the tools used to solve those problems.

Index

A
asthenosphere, 6, 10, 16
atmosphere, 4, 20, 25

B
biosphere, 4, 20

C
coal, 24, 26
continents, 11

E
Earth's age, 5
Earth's layers, 4, 6–7, 18
    crust, 6–7, 12, 20
    inner core, 6–7, 8, 9
    mantle, 6–7, 16
    outer core, 6–7, 8
earthquakes, 10, 12, 13, 18, 19, 22, 30
excavation, 22

F
fossil fuels, 26, 28, 29
fossils, 18

G
geosphere
    human interaction with, 22
    interactions with Earth's other systems, 4, 20
    protection of, 28–30
geothermal energy, 29

H
hydrosphere, 4, 20

I
igneous rock, 14, 16–17

L
landfills, 28
lava, 12, 20
lithosphere, 6–7, 10

M
magma, 12, 16
metamorphic rock, 14, 16–17
minerals, 4, 5, 19, 24

N
natural gas, 26
natural resources, 24, 26, 28, 30
nutrients, 24

P
Pompeii, 12

R
recycling, 28, 30
renewable energy sources, 28, 29
rock cycle, 14, 16–17
rocks, 4, 5, 12, 14, 16–17, 19, 24

S
sedimentary rock, 14, 16–17, 18
seismic waves, 18
Statue of Liberty, 25

T
tectonic plates, 7, 10–11, 12, 14

V
volcanoes, 10, 12, 16, 19, 20

Websites

Due to the changing nature of Internet links, PowerKids Press has developed an online list of websites related to the subject of this book. This site is updated regularly. Please use this link to access the list: www.powerkidslinks.com/ues/geo